HYDRAULIC FRACTURING IN THE MARCELLUS SHALE

Water and Health, Facts vs. Fiction

Written by
Theodore F. Them, MD, MS, PhD, MPH, FACOEM

A publication of the
ACSH American Council on Science and Health
Science. Not Hype.

American Council on Science and Health
1995 Broadway, Suite 202
New York, New York 10023-5860
Tel. (212) 362-7044. Fax (212) 362-4919
URL: http://www.acsh.org. Email: acsh@acsh.org

Publisher Name: American Council on Science and Health
Title: Hydraulic Fracturing in the Marcellus Shale: Water and Health, Facts Vs. Fiction
Price: $14.95
Author: Dr. Theodore F. Them
Subject (general): Science and Health
Publication Year: 2014
Binding Type (i.e. perfect (soft) or hardcover): Perfect

ISBN: 978-0-9910055-1-2

THE AMERICAN COUNCIL ON SCIENCE AND HEALTH APPRECIATES THE CONTRIBUTIONS OF THE FOLLOWING INDIVIDUALS WHO REVIEWED THIS REPORT

JILL KRIESKY, PH.D.
Associate Director of the Southwest Pennsylvania Environmental Health Project
McMurray, PA

ANTHONY J. GRIPPO, MD, MPH
Associate Physician, Section of Occupational Medicine, Guthrie Clinic, Ltd.
Sayre, PA

MICHAEL HOLLAND, MD, FACMY, FACOEM, FACEP
Toxicologist at Glen Falls Hospital
Glen Falls, NY

ISAAC ORR
Research fellow at The Heartland Institute
Chicago, IL

Acknowledgements. The paper relies on the work of numerous researchers in many fields, and these are referenced. A number of distinguished reviewers gave excellent comments and are gratefully acknowledged.

Conflict of interest. None declared.

Author information. Dr. Theodore F. Them is currently the Chief of the Section of Occupational and Environmental Medicine with Guthrie Medical Group in Sayre, Pennsylvania. He got his B.A. in Chemistry from Gettysburg College, his M.S. in chemistry from the University of New Mexico, his Ph.D. in theoretical inorganic and bioinorganic chemistry from the University of New Mexico and his MPH from the Medical College of Wisconsin. Throughout his professional career, Dr. Them has done research as an environmental analytical chemist specializing in forensic chemistry, but has worked in the area of environmental and occupational medicine since 1991. His special professional interests include environmental toxicology, workers'-compensation law and wellness programs.

TABLE OF CONTENTS

ABSTRACT

Increased awareness about using a highly technical process called hydraulic fracturing to recover natural gas trapped deep within the Marcellus shale has created questions about related human-health and environmental impacts. Associated arguments, both pro and con, have often been subjective, emotional, and unscientific. This paper is aimed at a systematic, objective review of documented types and rates of hydrofracturing-fluid- and chemical-related incidents affecting human health, to date, in the region of the Marcellus Shale Play.

1 THE MARCELLUS SHALE DEPOSIT

Also known as the Marcellus Shale Play, the Marcellus Formation, or, simply, "the Marcellus," this crescent-shaped, organic-rich, Middle Devonian shale deposit spans the (subsurface) distance from Albany, New York, through much of upstate New York; more than half of Pennsylvania (excluding mainly the southeastern area); small areas of northwestern New Jersey, northwestern Maryland, eastern Ohio, and western Virginia; virtually all of West Virginia; and relatively inconsequential parts of both eastern Kentucky and northeast Tennessee (see **Figure 1**, below).

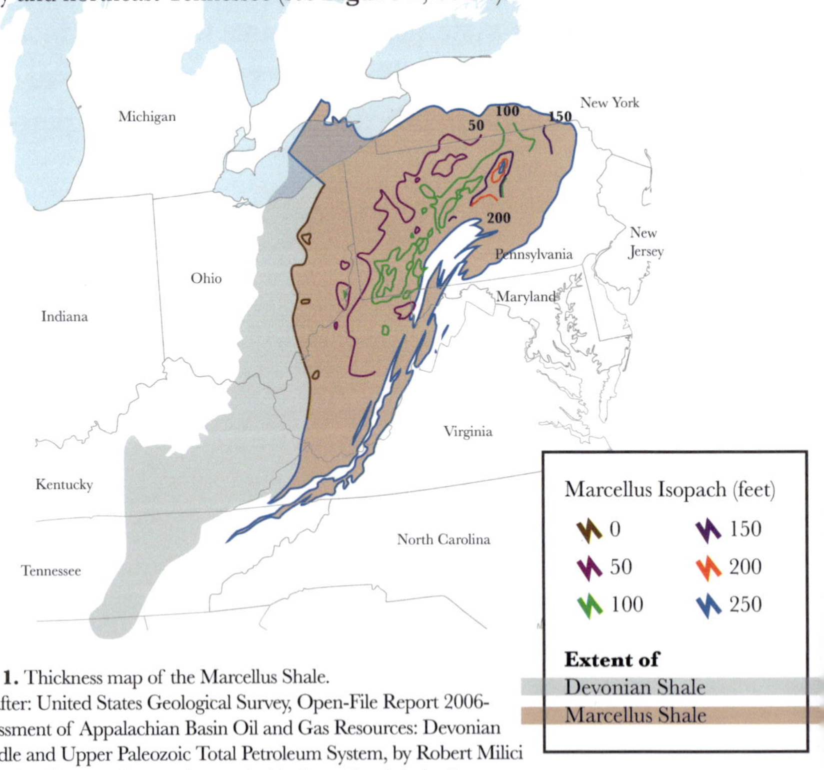

Figure 1. Thickness map of the Marcellus Shale.
Modified after: United States Geological Survey, Open-File Report 2006-1237, Assessment of Appalachian Basin Oil and Gas Resources: Devonian Shale-Middle and Upper Paleozoic Total Petroleum System, by Robert Milici and Christopher Swezey *(www.geology.com/articles/ marcellus-shale.shtml)*.

Figure 1 (previous page) shows the Marcellus shale to vary from only 50 feet to 250 feet in total thickness throughout its course.

Figure 2 (below) shows the approximate depth to the base of the Marcellus Shale throughout its course, that varying from 3,000-9,000 feet.

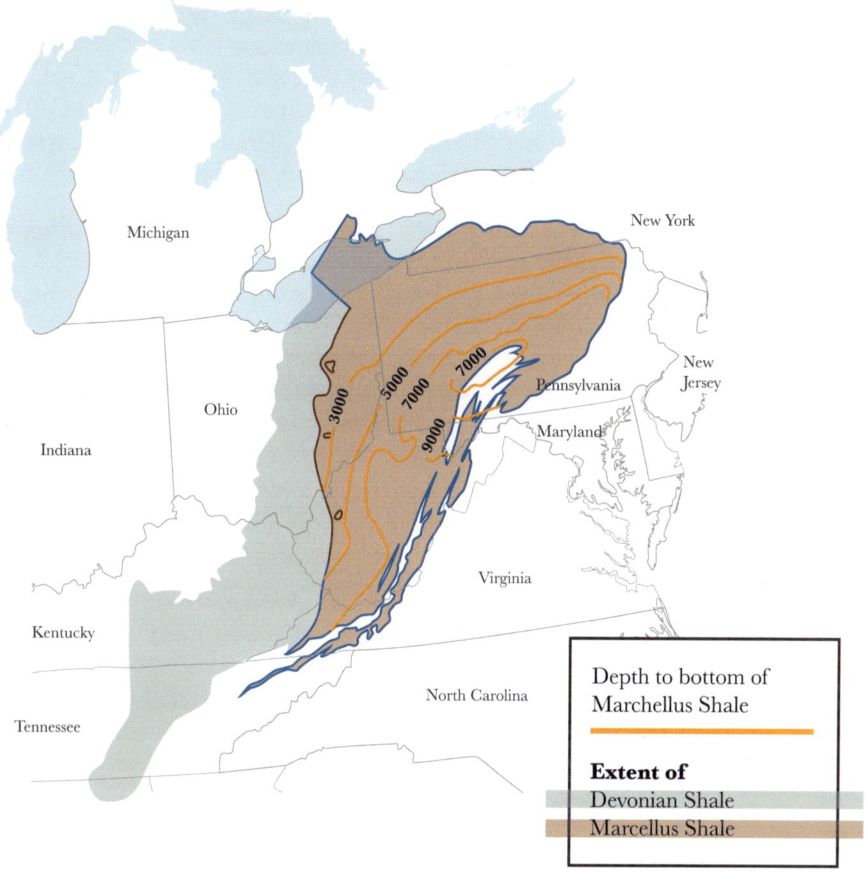

Figure 2. Prepared using the map by Robert Milici and Christopher Swezey above and adding depth-to-Marcellus contours published by Wallace de Witt and others. 1993, United States Department of Energy Report: The Atlas of Major Appalachian Gas Plays *(www.geology.com/articles/marcellus-shale.shtml)*.

The Marcellus Shale underlies terrain as varied as major metropolitan areas and vast stretches of rural, northeastern America. It is estimated to harbor more than 500 trillion cubic feet of natural gas, mainly methane, with an estimated wellhead value of $1 trillion.[1] That accounts for roughly one fifth (1/5) of all natural gas (recoverable) reserves in the entire U.S., which were estimated at 2,384 trillion cubic feet in 2012[2]. When compared against global natural gas reserves estimated at 6,707 trillion cubic feet[3] in 2011, the Marcellus reserves represent an impressive 7.5% of that total. At the current rate, the U.S. annually consumes roughly 25.5 trillion cubic feet of natural gas[4], meaning that the Marcellus alone could supply enough natural gas to meet U.S. needs for nearly 20 years. The 2,384 trillion cubic feet of total, U.S.-based natural-gas reserves, in turn, could supply, at present rates, this nation's natural-gas needs for roughly 93 years.

The Marcellus is characterized as a "tight" shale, with few naturally-occurring vertical fractures ("joints"), thereby currently necessitating the application of horizontal, subsurface drilling techniques, with engineering procedures that invoke subsequent, high-pressure, high-volume hydraulic fracturing (HVHF) and related techniques in order to make gas recovery both voluminous and financially feasible. Tight shales require fracturing because, although they are very porous, those pores are not interconnected and are also known as "not permeable" because of this. The fracturing process creates an artificial permeability by breaking these rocks apart and therefore allowing natural gas to flow up through the well. The lack of (natural) vertical joints is actually a huge benefit from a safety perspective in that vertical fractures would facilitate methane migration upward, toward the surface.

Typical HVHF fluids (HVHFFs) contain different chemicals of various types, manufacture, and concentration, including water (3-9M gallons/well, 9M being on the high side), sand (a "proppant," used to shore open the thousands of shale cracks resulting from the hydrofracturing process), biocides, rust/scale inhibitors, "acidizers" (acidifiers), other pH-adjusting agents, breakers, oxygen scavengers, crosslinkers, clay inhibitors, clay stabilizers, gelling agents, surfactants, and others. Some of these formulations remain proprietary; some have been made public.[5] Water and sand, together, comprise some 99.6% of typical HVHFFs, and the remaining chemicals just described comprise the remaining 0.4% of these typical HVHFFs. Representative, contributory fractions to that 0.4% multichemical content of the HVHFFs are seen in **Figure 3**, below.

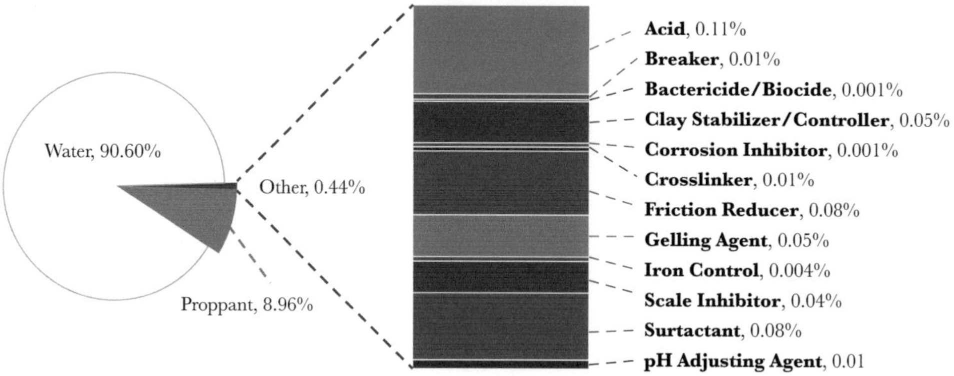

Figure 3. 'Sample Fracture Fluid Composition by Weight.' As seen in the draft Supplemental Generic Environmental Impact Statement on natural gas drilling by the New York Department of Environmental Conservation Division of Mineral Resources *(www.propublica.org/feature/gas-execs-call-for-disclosure-of-chemicals-used-in-hydraulic-fracturing-102).*

Despite the routine presence of similar, or identical, chemicals in many American foodstuffs, medications, cosmetics, and household cleaners[6] in much higher concentrations, much has been published about potential environmental and human-health impact from the latter components of the HVHFFs.[7] (There are literally hundreds of similar publications, easily found on the Internet; however, the cited source serves to exemplify, in general, this phenomenon.) However, most of these latter types of references are devoid of critical analyses and discussions of dose-response relationships so germane to related toxicity arguments. Moreover, to date, relatively little has been published regarding *actual* Marcellus-related HVHFF-related incidents, their natures, their numbers, their geographic locations, their rates, their verifications, and/or their true (not imagined, hypothesized, or theorized) ultimate impact(s) along these lines.

The hydrofracturing process is associated with both natural and engineered barriers to natural-gas flow, leakage, and contamination from the Marcellus wells, including distance/depth, steel and cement/steel well casings, blow-out preventers, pressure sensors, and many complex series of control valves and manifolds. A related diagram of the overall structural setup in a typical Marcellus well is seen in **Figure 4**, next page. Chemical-transportation controls, specifically, involve appropriately rated vehicles, the routine use of federally mandated vehicle placarding, and controlled volumes of HVHFF chemicals. Wellsite surface-safety controls require controlled access/security, chemical isolation/protection, and personal protective equipment (PPE) for nearby workers of many trades. Surface controls at the wellhead also include pressure-rated plumbing and hardware; pressure and flow sensors/alarms; impermeable surface/ground mats; mixing and injecting controllers; holding/waste ponds; recovering/recycling; and processing of HVHFF chemicals. In the active-drilling Marcellus zones, flowback water is subject to processing/cleanup prior to discharge, and as much as 90% of that flowback water is re-used at different well sites.

Figure 4. Typical wellsite engineering at a Marcellus hydrofracturing site *(www.anadarko.com/SiteCollectionDocuments/PDF/AXPC%20Real%20Facts%20about%20FracStim(2)012710.pdf)*.

Down-hole safety measures concerning HVHFFs also include pressure-rated plumbing and hardware; well casings (typically now triple-steel/cement in the area of groundwater aquifers and the vertical component of the well shafts); blowout preventers; flow/position/pressure monitors and sensors; and simple geologic depth. Despite these numerous safeguards, there remains a great deal of public skepticism about their reliability and effectiveness in preventing HVHFF contamination of wellsite surfaces, groundwater, air, and surface water. ■

2 STUDY DESIGN AND TECHNIQUES

This is an observational study whose goal is to objectively characterize and quantify any and all subsurface, HVHF-related incidents that, to date, have actually adversely affected human health in the distribution of natural-gas drilling and production activities in the Marcellus. This study is restricted to the actual hydrofracturing process itself, not to the surface activities of drilling; the subsurface activities of drilling; the building of related roads or pipelines; or the processes of casing a well – all of which are common to drilling of any well for any purpose, including at least 15 million domestic water wells[8] (2007 data) and at least 825,000 active oil and gas wells[9] (2009 data) drilled in the U.S. since 1821, and the 2.5 million total wells (including both active and inactive oil and gas) drilled in the U.S. since 1950.[10] The New York State Department of Environmental Conservation has independently endorsed the essential equivalence of HVHF and other types of drilling, stating, in its own words, "Except for special tools used underground, horizontal drilling is performed using the same equipment and technology as vertical drilling, with the same protocols in place for aquifer protection, fluid containment, and waste handling."[11] This study also only concerns itself with documented, adverse human-health impact to date, not future impact(s) of any sort.

The nine states overlying the Marcellus Shale deposit (Kentucky, Maryland, New Jersey, New York, Ohio, Pennsylvania, Tennessee, Virginia, and West Virginia) were queried, in writing, by way of their respective Departments of Health (DOH), Departments of Environmental Protection/Conservation (DEP or DEC), on March 07, 2013, as well as via their respective U.S. Environmental Protection Agency (US EPA) Regional Offices and the US EPA headquarters in Washington, D.C., (on April 18, 2013) with respect to several key pieces of data. Those requests, placed with these various agencies under applicable state and federal freedom-of-information legislation, were for the following and were related only to Marcellus Shale hydrofracturing activities:

1. Investigations and related summaries, letters, memoranda, reports, findings, and outcomes of any surface (above-ground) incident(s) involving hydrofracturing fluids and/or their constituent chemicals in which human health and/or the local environment was, or may have been, adversely impacted, listed by county, city/town/township/location, date, and time.

2. Investigations and related summaries, letters, memoranda, reports, findings, and outcomes of any down-hole (below-ground) incident(s) involving hydrofracturing fluids and/or their constituent chemicals in which human health and/or the local environment was, or may have been, adversely impacted, listed by county, city/town/township/location, date, and time.

3. Investigations and related summaries, letters, memoranda, reports, findings, and outcomes of any subsurface (below-ground) or ground-water incident(s) involving hydrofracturing fluids and/or their constituent chemicals in which human health and/or the local environment was, or may have been, adversely impacted, listed by county, city/town/township/location, date, and time.

4. Precise number of gas wells permitted, as well as the actual number of gas wells drilled, listed by county, city/town/township/location, and date.

Additionally, as the above-requested, written information was received, other, secondary, private-, state-, and federal-level sources of germane data were identified; those were pursued toward similar ends and were included in this assessment. ■

3 STUDY RESULTS

Table 1 summarizes the results received from each respective state's Department of Health (DOH) and Department of Environmental Conservation/Protection (DEC/P) and from the US EPA Regional Office overseeing that state (US EPA). (Note: All references to human-health impact(s) hereinafter specifically refer to adverse ones, not otherwise.)

STATE ⟶ AGENCY, RESPONSE, AND DATE OF RESPONSE ↓

	KY	MD	NJ	NY
DOH	NRM[1*] and no actual drilling/ HVHF activity in state **3/26/13**	NDP[2*] and no actual drilling/HVHF activity in state **3/27/13**	NDP, with no actual drilling/HVHF activity in state **3/26/13**	RTO[3*] 5/28/13 and NFR[4*] through **11/01/13**
DEC/P	NDP, with claims that the Marcellus Shale is not present in KY and no actual drilling/HVHF activity in state **3/25/13**	NDP and RTO and no actual drilling/ HVHF activity in state **3/21/13**	NDP and no actual drilling/HVHF activity in state **4/02/13**	NDP on items 1-3 of request and RTO re item 4 6/12/13 and NDP and RTO **1/21/14**
US EPA, Regional Office	NDP, Region[4] **4/23/13**	Region 3: Approx 40[6*] incidents involving spills into navigable waters[6] and NDP directly regarding HVHF activity **9/17/13**	Region 2: No records release without paid fee; fee waiver denied NA[7*]	Region 2: No records release without paid fee; fee waiver denied NA[7*]

Table 1. Primary state- and federal-based responses to written inquiries cited above. (DOH – Department of Health; DEC/P – Department of Environmental Conservation/Protection; US EPA – United States Environmental Protection Agency, by region).

[1]*NRM – none related to the Marcellus Shale

[2]*NDP – no documents pertaining/responsive to the questions posed

[3]*RTO – referred to other database or information source

[4]*NFR – no final response received to date (inquiry receipt, however, was acknowledged)

[5]*NAI – not aware of any such incident(s).

[6]*Verbal communication with US EPA, Region 3, April 19, 2013, reflecting US EPA's involvement in these cases, none with documented human-health impact. The approximately 40 incidents in US EPA Region 3 reflect the total number of incidents within the four states listed above, which fall into US EPA Region 3 here (MD, PA, VA, and WV), without further specifics.

[7]*NA – not applicable; no Marcellus drilling in NJ or NY

OH	PA	TN	VA	WV
NDP and RTO **3/28/13**	NFR **3/21/13** (last response)	NAI[5]* and RTO and no actual drilling/HVHF activity in state **3/26/13**	NDP and RTO **3/19/13**	NDP and RTO **3/19/13**
NFR, with indirect response from OH DOH **3/28/13**	Teleconference and RTO **3/28/13**	NAI and RTO, with claims that the Marcellus Shale is not present in TN **3/16/13**	NDP, with no actual drilling/ HVHF activity in state **3/20/13**	NFR and RTO, with indirect response from the WV Dept. of Env. Health Svcs. **3/19/13**
NDP and RTO, Region 5 **4/03/13**	Region 3: Approx 40 incidents involving spills into navigable waters[6]* and NDP directly regarding HVHF activity **9/17/13**	NDP, Region 4 **4/23/13**	Region 3: Approx 40 incidents involving spills into navigable waters[6]* and NDP directly regarding HVHF activity **9/17/13**	Region 3: Approx 40 incidents involving spills into navigable waters[6]* and NDP directly regarding HVHF activity **9/17/13**

In short, and as shown in **Table 1** (above), five (5) of the nine (9) potentially affected states – namely, Kentucky, Maryland, New Jersey, Tennessee, and Virginia – as of this writing have not experienced any drilling activity or, therefore, any related HVHFF use or incidents in the Marcellus. New York (before the current drilling ban on HVHF (see page 25), Ohio, Pennsylvania, and West Virginia, on the other hand, have experienced such activity, albeit with no confirmed incidents proven to have impacted human health in the latter three states. US EPA Region 2, covering New York has, to date, not provided any final response(s) to the questions posed; a fee-waiver appeal to US EPA headquarters in Washington, D.C., regarding associated information-release costs has been denied as of this writing. The individual US EPA Regional Offices where there is current Marcellus activity, however, did respond to the inquiries, and resultant data are listed below.

Pursuant to receipt and analysis of the information cited in **Table 1** (above), formally recommended, secondary sources of potentially pertinent information were pursued. Those sources are summarized in **Table 2** (below). Maryland, New Jersey, Tennessee, and Virginia have been excluded from **Table 2** due to lack of related drilling activity in those states. (When this Table was created, it was not yet evident that there was no related drilling activity in Kentucky.)

STATE OR US EPA REGION ↓ SOURCE →

	Internet	Other Agency	Other Individual
KY	None	Kentucky Division of Oil and Gas	None
NY	None	None	None
OH	None	Ohio Department of Natural Resources	None
PA	PA DEP eFacts *(http://www.ahs.dep.pa.gov/eFACTSWeb/)*	None	Ms. Laura Legere, Journalist, The Scranton Times-Tribune, 149 Penn Ave., Scranton, PA 18503 Phone: 570-349-9100 (No longer employed there.) *http://thetimes-tribune.com/news/sunday-times-review-of-dep-drilling-records-reveals-water-damage-murky-testing-methods-1.1491547* and *http://thetimes-tribune.com/news/gas-drilling-complaints-map-1.1490926* and *http://maps.fractracker.org/2.0/?appid=8833c19d11374920ac1cec6b6c60d497*
WV	None	None	None
US EPA Region 3	*http://www.nrc.uscg.mil/foia.html*		

Table 2. Secondary information sources resulting from primary contacts listed in Table 1, above.

Table 3 (below) summarizes the findings of the Pennsylvania Department of Environmental Protection's (PA DEP's) formal findings in the matter of 973 separate properties that were either sent letters or were named in consent orders between the start of 2008 and fall 2012 (roughly September). Unfortunately, the letters are not necessarily very specific about what caused a disruption to the cited water supplies. What one can find in the spreadsheet in **Table 2** (database collected by Laura Legere reference), 969 cases, and four (4) additionally, recently reported cases described in a private communication of August 12, 2013, with the author of that spreadsheet, under the "causality established" field, is whether or not PA DEP made a final determination that oil and gas operations of any kind were to blame for either diminishing the water flow or polluting the water source.

PA DEP's Findings in 973 Cases of Interest →	**Confirmed Pollution or Diminution[1#] by Drilling Operations**	**Unclear or Unfinished at Time of Writing**	**No Drilling-Related Impact**
Number of Cases →	162[2#]	69[3#]	742

Table 3. Pennsylvania Department of Environmental Protection's (PA DEP's) findings in 969 cases of interest.

[1#]"Diminution" is meant to define otherwise-inexplicable drawdown on, or reduction in, local groundwater supplies.

[2#]Latest, single case reported actually precedes the period of interest (2008-2013) by 3 years, at which time there was no drilling or HVHF activity in the Marcellus.

[3#]Three (3) of these cases were reported after the original publication *(https://dl.dropboxusercontent.com/u /48182083/ScrantonTT_GW_RTK.csv)* cited above and remained undetermined as of this writing.

Table 4 (below) further details PA DEP's formal findings, as seen in the spreadsheet cited in **Table 2**.[12]

PA DEP'S FORMAL INVESTIGATIVE CATEGORIES AND NUMBERS IN EACH CATEGORY⟶
PA DEP'S DETAILED FINDINGS ↓

	Diminution[1^] Complaints: 102	Combined Pollution/Diminution Complaints:41	Pollution Complaints: 830
Attributed to Oil and Gas Operations	5	6	151
Not Attributed to Oil and Gas Operations	96	35	611
Undetermined	1[2^]	0	68

Table 4. Pennsylvania Department of Environmental Protection's (PA DEP's) investigative categories, related numbers, and detailed findings.

[1^] "Diminution" is meant to define otherwise-inexplicable drawdown on, or reduction in, local groundwater supplies.

[2^] Although "undetermined" regarding causation, the oil and gas company in question, according to the sources cited above, drilled a new water well for the resident-complainant.

Table 5 (below) further elucidates detailed findings by the PA DEP regarding cases in which it determined that a water supply was, indeed, contaminated by oil or gas activities, as listed in the spreadsheet described above.[12]

PA DEP's Formal Findings and Categories Regarding 156 Cases of Water Supplies Contaminated by Oil and Gas Activities ↓	Numbers of Cases ↓
Elevated Methane Alone or Methane Plus Other Elevated Parameters (TDS [total dissolved solids], barium, manganese, iron, etc.).	89
Impact, Without Identification of Specific Cause	23
Elevated Levels or Concentrations of TDS (total dissolved solids), Chloride, Sodium, Manganese, Iron, Barium, Strontium, Turbidity and/or Aluminum	43
Oil	1[1@]

Table 5. Pennsylvania Department of Environmental Protection's (PA DEP's) formal findings in 156 cases of what it determined were water supplies contaminated by oil and gas activities. With respect to column 2, above, it is not clear as to whether or not the PA DEP differentiated, or even tried to differentiate, between biogenic methane or thermogenic (fossil-sourced) methane species.

[1@]Apparently sourced from a nearby, abandoned well.

The PA DEP's letters in these matters[12] almost never cited a specific cause of contamination, but when they did, the causes were not typically due to underground HVHF activities; they were typically surface phenomena, as shown in **Table 6** (below).

PA DEP's Causation Categorizations Regarding Water-Supply Contamination ↓	Numbers of Homes or Springs Affected ↓
Drilling and Drilling-Related Road Construction	2 Homes
Impoundment Construction	1 Home
Pipeline Construction	2 Homes
Stray Cement or Cuttings	7 Homes
Brine Contamination[1+]	6 Springs
Hydraulic Fracturing, Directly	0 Homes, 0 Springs

Table 6. Pennsylvania Department of Environmental Protection's (PA DEP's) causation categorizations regarding water-supply contamination related to Marcellus operations.

[1+]Suggesting, per PA DEP, "an infiltration of frac water into the shallow ground water," generally from a "surface discharge."

Of note, per the author of spreadsheet cited in Table 2, which lead to **Table 6** (above), "none of the letters and orders I reviewed ever noted contamination caused by hydraulic fracturing directly"[13] (emphasis added).

Table 7 (below) summarizes the confirmed numbers and types of incidents, in total, by state, discovered during the review of all inquiry-related responses shown in **Table 1** (above) and all secondary sources of information shown in **Table 2** (above). Kentucky, Maryland, New Jersey, Tennessee, and Virginia have been excluded from **Table 7** due to lack of related drilling activity in those states.

STATE ↓ NUMBERS AND TYPES OF INCIDENTS →

	Environmental, all types	Environmental, with Documented Human-Health Impact
NY	NFR[1**]	NFR
OH	0	0
PA	All of US EPA Region 3: total of ¬40 spills into navigable waters. PA DEP[2**]: Total of 162	US EPA Region 3: 0 PA DEP: 0
WV	All of US EPA Region 3: total of ¬40 "spills into navigable waters". WV DHHR/BPH/OEHS[3**]: 0	US EPA Region 3: 0 WV DHHR/BPH/OEHS: 0

Table 7. Confirmed numbers and types of HVHF-related incidents by state, as derived from both primary and secondary information sources described in Tables 1 and 2 (above), respectively.

[1**]NFR – no final response received to date (inquiry receipt, however, was acknowledged).

[2**]PA DEP – Pennsylvania Department of Environmental Protection.

[3**]WV DHHR/BPH – West Virginia Department of Health and Human Resources/Bureau of Public Health/Office of Environmental Health Services.

Regarding the approximately 40 incidents of "spills into navigable waters" cited by US EPA Region 3 (see **Table 2**, above), which covers Delaware, the District of Columbia, Maryland, Pennsylvania, Virginia, and West Virginia, this author was unable to confirm each and every one of those incidents from the referenced web site. For example, some of the citations in the U.S. Coast Guard's related web site described reports of individuals who were "worried" about local hydrofracturing activities or who reported oil leaking from a gasfield-affiliated truck onto a neighbor's driveway or into the neighboring street – incidents that may reflect indirect, but not direct, relationships to HVHF activities. Given this, it is clear that what is/was characterized as "spills into navigable waters" does not necessarily constitute, as just stated, activities linked *directly* to the hydrofracturing process itself and were not further considered.

Moreover, while the US EPA Headquarters was unwilling to surrender related documents without receipt of some potentially prohibitive fees, even after a formal and related fee-waiver appeal, each of the independently queried Regional Offices of concern, except Regions 2, answered this study's inquiries with a uniform answer of "no data responsive" to such inquiries. Region 2 of the US EPA, like US EPA Headquarters, required a fee for release of related records; however, neither NY nor NJ has active Marcellus drilling. These collective facts render the related information kept at both US EPA Region 2 and US EPA Headquarters in Washington, D.C., moot. The information eventually gleaned from US EPA Region 3 regarding the hydrofracturing process itself, was, in turn, "no data responsive" to the inquiry placed through that office. (This author wishes to note the cooperative and informative nature of all communications, both written and verbal, with US EPA Region 3 in this process; assigned personnel there were eager to provide accurate information and did so in a surprisingly rapid, professional, and comprehensive manner, with appropriate follow up.)

Table 8 (below) shows the actual number of natural gas wells drilled, as of this writing, using HVHF, state by state, in the Marcellus Shale, with related references.

State ↓	**Number of Wells ↓**
KY	0
MD	0
NJ	0
NY	NFR[1!] from the NY DOH; NY DEC answered in part, with the statement described below (page 25) but released no final, conclusive data
OH	460[2!]
PA	8,982 to 9,848[3!]
TN	0
VA	0
WV	approximately 1100[4!]

Table 8. Number of natural-gas wells drilled, as of this writing, using HVHF, state by state, in the Marcellus Shale, with related references.

[1!]NFR – no final response received to date (inquiry receipt, however, was acknowledged).

[2!]See *marcellusdrilling.com/2013/08/latest-oh-utica-numbers-819-permits-460-wells-drilled/*. (Caveat: This appears to be a combined number for both Marcellus and Utica shale plays in Ohio.)

[3!]See *stateimpact.npr.org/pennsylvania/tag/ marcellus-locations-app/ and www.ohio.com/blogs/drilling/ohio-utica-shale-1.291290/ pennsylvania-has-9-848-marcellus-shale-wells-1.310258.*

[4!]Personal communication, Thomas J. Aluise, West Virginia Department of Environmental Protection, August 8, 2013.

Table 9 (below) shows the incidence (rate) of environmental incidents and environmental incidents with documented human-health impact directly due to HVHF activity, per 100 wells drilled in the Marcellus Shale, using HVHF technology, in each state. These data derive from above-cited information and related calculations. ■

STATE → INCIDENCE OF VARIOUS TYPES OF INCIDENTS (PER 100 MARCELLUS/HVHF WELLS DRILLED) ↓

	KY	MD	NJ	NY	OH	PA	TN	VA	WV
Environmental	NA[1^^]	NA	NA	NFFR[2^^]	0	1.64-1.80	NA	NA	0
Environmental, with Documented Human-Health Impact Directly Due to HVHF Activity	NA	NA	NA	NFFR	0	0	NA	NA	0

Table 9. Incidence (rate) of environmental incidents and environmental incidents with documented human-health impact, per 100 wells drilled in the Marcellus Shale, using HVHF technology, in each state.

[1^^]NA - not applicable to this state; no Marcellus wells drilled.

[2^^]NFFR – no final figures received as of this writing.

4 DISCUSSION

One might reasonably question the propriety of contacting the US EPA and the respective states' Departments of Health and Departments of Environmental Protection/Conservation for health information related to this matter. That is to say, were these the correct agencies to query with respect to human-health data related to drilling/production in the Marcellus? That answer may be objectively gleaned from the respective mission statements of these agencies.

The US EPA lists its mission as the following.

The mission of EPA is to ***protect human health*** *and the environment.*

EPA's purpose is to ensure that:

- *all Americans are protected from significant risks to* ***human health*** *and the environment where they live, learn and work;*
- *national efforts to reduce environmental risk are based on the best available scientific information;*
- *federal laws protecting* ***human health*** *and the environment are enforced fairly and effectively;*
- *environmental protection is an integral consideration in U.S. policies concerning natural resources,* ***human health****, economic growth, energy, transportation, agriculture, industry, and international trade, and these factors are similarly considered in establishing environmental policy;*
- *all parts of society — communities, individuals, businesses, and state, local and tribal governments — have access to accurate information sufficient to effectively participate in managing* ***human health*** *and environmental risks;*
- *environmental protection contributes to making our communities and ecosystems diverse, sustainable and economically productive; and*
- *the United States plays a leadership role in working with other nations to protect the global environment*[14] *(emphasis added).*

Of the nine states overlying the Marcellus Shale, five (Kentucky, Maryland, New Jersey, Tennessee, and Virginia) have not experienced any related drilling to date. Of the remaining four states, only three admit to active drilling with HVHFFs in the Marcellus Shale, namely Pennsylvania, Ohio, and West Virginia. (New York continues to permit low-volume, but not high-volume, hydrofracturing; see page 25.) Mission statements for the Departments of Health and Departments of Environmental Protection/Conservation for the latter, three states are as follows.

The Pennsylvania Department of Health's related statement is:

The mission of PA DOH is to promote healthy lifestyles, **prevent injuries and disease**, and to ensure the safe delivery of quality healthcare for all Commonwealth citizens[15] (emphasis added).

The Pennsylvania Department of Environmental Protection lists its mission statement as:

The Department of Environmental Protection's mission is to protect Pennsylvania's air, land and water from pollution and to ***provide for the health and safety*** *of its citizens through a cleaner environment. We will work as partners with individuals, organizations, governments and businesses to prevent pollution and restore our natural resources*[16] *(emphasis added).*

The Ohio Department of Health similarly states the matter as:

The Ohio Department of Health's (ODH) mission is to ***protect and improve the health*** *of all Ohioans by* ***preventing disease****, promoting good health, and assuring access to quality health care. To accomplish that goal, it is important that we consistently regulate Ohio's public health structure*[17] *(emphasis added).*

And, the Ohio Environmental Protection Agency's related statement is:

Our mission is to ***protect*** *the environment and* ***public health*** *by ensuring compliance with environmental laws and demonstrating leadership in environmental stewardship*[18] *(emphasis added).*

West Virginia's Department of Health and Human Resources statement is:

The Department of Health and Human Resources' mission is to promote and provide ***appropriate health and human services*** *for the people of West Virginia in order to improve their quality of life*[19] *(emphasis added).*

Finally, West Virginia's Department of Environmental Protection's statement is:

The West Virginia Department of Environmental Protection's mission is to promote a ***healthy*** *environment*[20] *(emphasis added).*

These mission statements strongly suggest that these respective agencies were, indeed, the proper ones to contact and to query, as described above. While the remaining, respective Departments of Health and Departments of Environmental Protection/Conservation in the six (6) states (Kentucky, Maryland, New Jersey, New York, Tennessee, and Virginia) overlying the Marcellus shale post similar mission statements, which are easily found online, these are rendered moot with respect to related queries in the current absence of Marcellus HVHF drilling/ production activities in those states.

The State of New York's Department of Health (DOH) has not, to date, provided a final response to the author's related inquiries despite repeated, monthly, documented commitments from that State to do so. Accordingly, there are no direct data from the New York DOH to include in this study. There is currently a statewide ban on *H*VHF (with emphasis on the "H") in New York since 2008, imposed by Governor Cuomo, which is especially curious in light of the fact that, as published by the New York Department of Environmental Conservation itself, "Horizontal drilling has been used in New York since the 1980s."[21] The latter (horizontal-drilling activity in New York) apparently spanned nearly three decades (or more; see below). The author also requested information similar to that otherwise requested, as cited above, from the New York Department of Health with respect to any documented adverse human-health events from the horizontal drilling that had been completed there "since the 1980s," but, once again, New York DOH failed to provide any final, formal response whatsoever to that request to date. An extensive Internet search by the author has failed to yield any confirmed, documented impact on human health in New York from three or more decades of "horizontal drilling" there, some of which *must* have involved hydrofracturing. (The author has spoken directly with one seasoned well driller from New York who personally revealed that he had been "hydrofracturing in New York for over 25 years, with no problems at all.")

A recent, repeat inquiry on this matter to the New York Department of Environmental Conservation has generated a rather stunning revelation, quoted verbatim below.

- *the current "hydrofracturing ban" imposed in 2008 is for high volume hydraulic fracturing (HVHF) which involves stimulation with greater than 80,000 gallons of water*
- *low volume stimulations have not been banned and have been occurring in the state for approximately 50 to 60 years (perhaps this is what the driller was trying to convey to you)*
- *DEC estimates that 90% of all gas wells drilled in the state required low volume fracture stimulation in an attempt to commercially produce the wells*
- *the 75,000 wells drilled stat includes many well types other than gas wells (90% times 75,000 would yield an incorrect gas well stimulation total)*

We cannot provide the answer to your question. I recommend that you file a FOIL request if the above information does not meet your needs.[22]

Given this, it is evident that – despite a so-called "ban" on hydrofracturing in New York, both the techniques and technologies of hydrofracturing have remained ongoing to date, albeit with smaller *volumes* (not necessarily *types*) of hydrofracturing fluids. Despite this author's attempts to secure related human-health data from both the New York Department of Health and the New York Department of Environmental Conservation, neither Department ultimately provided such data. Given, though, that hydrofracturing in New York has not only been ongoing there for "50 to 60 years,"[22] and remains ongoing to date (with volumes less than 80,000 gallons), such horizontal drilling in New York over those several decades appears to have maintained without any evident and/or traceable adverse human-health impact(s); otherwise, one would have expected well publicized epidemiologic reports and traceable human-health data on such matters, with a much earlier-imposed, comprehensive ban on any/all types and volumes of hydrofracturing fluids. That is an implied safety record that has apparently been overlooked by that State with respect to its present ban on the same *type* of drilling, the only difference apparently being the volumes of hydrofracturing fluids currently used to accomplish the same end result. It is apparently only the absolute *volumes* of hydrofracturing fluids in HVHF to which New York objects, not the actual *technique* or *technology* of that process. The author physically resides in an active Marcellus-drilling zone adjacent to several New York counties that overlie the Marcellus as well, and it most certainly is *not* the author's experience or opinion that the anti-hydrofracturing activists in, or residents of, those neighboring New York counties are fully aware of this matter.

Of the three states with active Marcellus HVHF drilling, two of them, namely Ohio and West Virginia, have not experienced any officially documented, adverse human-health effects directly related to that activity, even with hundreds of wells already having been drilled there.

Pennsylvania, the state with the highest level of HVHF activity in the Marcellus, with roughly 10,000 wells drilled there from 2008 to date, has experienced roughly 40 related environmental incidents ("spills into navigable waters") in which the US EPA became involved via its Region 3 office, none of which apparently rose to the level of adverse human-health events. The US EPA has become embroiled in many cases of suspected, Marcellus-related environmental disruption, including the infamous one of alleged, widespread groundwater contamination of domestic wells by migrating methane in Dimock, Pennsylvania[23,24]; however, the US EPA closed that investigation, with findings that the level of contamination was below federal safety triggers.[25] Similarly, the US EPA has either terminated and/or otherwise significantly revised its investigations into several other allegedly HVHF-related incidents in many other states/shale plays.[25]

With respect to the study at hand, the EPA's total tallies on alleged, Marcellus-/HVHF-related environmental infractions were sourced from the US EPA Region 3's verbal representations to this author of approximately forty (40) incidents of "spills into navigable waters" that had risen to its level of review. US EPA Region 3 holds purview over several states of related interest, including Maryland (no Marcellus drilling), Pennsylvania (roughly 10,000 Marcellus wells), Virginia (no Marcellus drilling), and West Virginia (roughly 1,100 Marcellus wells). Ohio, where the remainder of the Marcellus-drilling activity is located (roughly 460 Marcellus wells), falls in US EPA Region 5, in which there was no report of US EPA involvement in any related incident(s). While this represents a total US EPA involvement incidence of roughly 40/(10,000 + 1,100 + 460), or 40/11,560, or 0.35%, in the two active US EPA Regions (3 and 5), US EPA Region 3 has also confirmed "no data responsive" to the formal inquiry. Apparently, *none* of those 40 incidents of "spills into navigable waters" rose to the level of documented human-health impact for the roughly 11,100 Marcellus wells in its jurisdiction. US EPA Region 2, which oversees related activities in NY and NJ, where there is no current Marcellus activity, would not release related information without potentially prohibitive, associated fees. In the absence of Marcellus activity in NY and NJ, though, no related human-health impact(s) would be expected. Accordingly, absence of that information did not adversely affect the conclusions of this paper. The remainder of the states overlying the Marcellus, including Kentucky, Tennessee, and Virginia, have not experienced any Marcellus-related drilling and, therefore, have experienced no HVHF-related, documented environmental or (adverse) human-health impact(s), as suggested by their respective US EPA Regional Offices' responses of "no data responsive" to the formal inquiries of this study.

The Pennsylvania Department of Environmental Protection (PA DEP), according to reports published in the *Scranton Times-Tribune* and by *FrackTracker.com* (see **Table 2**, above), has been very actively investigating claims of alleged domestic water-well and surface-spring contamination from HVHF-related drilling, as cited herein. Of the 973 complaints along these lines, *none* was determined to be *directly* causally related to HVHF activity. Many of the PA DEP's cited data points lack precise dates (redacted), making it impossible to analyze such data for seasonal and/or rainfall-induced effects of potential interest.

While the presence of methane in domestic water was, in some cases, determined by the PA DEP to be causally related to "contamination of water supplies from oil and gas operations," in others it was *not* found by the PA DEP to be causally related to such "oil and gas operations." Due to the PA DEP's failure to respond to related inquiries on this matter, the precise methodologies used by the PA DEP in making its related determinations on methane in water supplies is not known to the author. Methane itself is virtually non-toxic to humans – and is actually produced in the gastrointestinal tracts of roughly 56% of humans studied[26] – but does represent a potential explosion hazard at concentrations of 5-15% in air, especially if entrapped.[27] While there have been sporadic reports of findings of such levels of methane in homes near Marcellus-drilling activities, there have been no citations of related explosions or injuries to humans or human health from this. Moreover, given the recent findings of methane in well water in 15% of 66 domestic wells tested by the U.S. Geological Service in the immediately neighboring upstate/Southern Tier area of New York[28] (which directly overlies the Marcellus shale) in the complete absence of related drilling *and* in the presence of a related drilling ban, the question arises as to whether such findings result from Marcellus drilling or simply from naturally occurring methane migration from underground sources. Those (New York) wells were more than one mile away from existing or abandoned natural gas wells, and the levels of methane were "high enough to warrant monitoring or remediation."[28] Four of those wells showed methane levels so high that "water coming out of a tap could potentially be lit with a match, or be an explosive risk."[28] There is no valid, scientific reason to believe that the same phenomenon is not occurring in Pennsylvania, as discussed elsewhere, above.

Furthermore, the PA DEP has made similar findings with respect to causation related to oil and gas operations regarding evidence of "TDS (total dissolved solids), chloride, sodium, manganese, iron, barium, strontium, turbidity and/or aluminum" in potentially affected water supplies. These species are *all* potential "background," or naturally occurring, ones, commonly seen in Pennsylvania groundwater. The absolute and relative concentrations of these chemical species may be subject to seasonal/rainfall fluctuations and/or other influences. (Research is ongoing on the latter matters as this is being written.) It is not evident that PA DEP, in its related findings, addressed such possibilities. Moreover, it is not clear as to whether the PA DEP performed the necessary radio-isotopic analyses so critical to differentiating thermogenic (fossil-fuel-sourced) methane from biogenic methane when making its related determinations on causation. (The author is unable to discern that matter based on the collective materials reviewed.)

Despite these numerous findings by the Pennsylvania Department of Environmental Protection, its sister organization, the Pennsylvania Department of Health (PA DOH) did not respond to the author's request for its official findings of any/all adverse human-health effects directly related to HVHF activity in the Marcellus. Secondary sources for related information, cited above, however, provided ample documentation on such matters in Pennsylvania.

The Departments of Health and the Departments of Environmental Protection/Conservation in the two remaining states (Ohio and West Virginia) in which there is, or has been, Marcellus-/ HVHF-drilling activity, submitted no documented evidence of adverse human-health impact resulting from hydrofracturing itself.

Finally, as this is being written, at least two University systems, namely the Pennsylvania State University and Duke University, are actively engaged in conducting studies on potential, Marcellus-related groundwater[29,30] and related human-health phenomena, mainly in Pennsylvania. Those studies are incomplete but will certainly bear review upon final publication. ■

5 CONCLUSIONS

The appropriate federal and state agencies were queried, as described above.

To date, there has been no Marcellus-related HVHF activity in Kentucky, Maryland, New Jersey, Tennessee, and Virginia; therefore, there is no related human-health impact.

New York, which overlies the Marcellus, has apparently experienced historical/pre-ban HVHF activity since the 1980s, but neither of New York's queried agencies (DOH and DEC) has provided final, requested data to this author to date, thereby currently preventing calculations of officially recorded incidence of HVHF-related human-health impacts from those historic activities in that state, by its cited agencies. Despite this, given the absence of post-ban HVHF-type drilling in New York, one would not logically expect any such impacts. (While New York apparently continues to permit low-volume hydrofracturing [LVHF], that is not the thrust of this paper.) Beyond that, extensive Internet searches failed to produce any objective, traceable data on adverse human-health impacts from "horizontal drilling" and/or hydrofracturing in either the pre-ban (HVHF *and* LVHF permitted) and/or post-ban (LVHF *only*) eras in New York.

Similarly, the US EPA's Region 2, which oversees New York, did not provide requested data on any such incidents. In the absence of, or ban on, recent Marcellus-related, HVHF-sourced drilling there, however, it is, once again, likely that there are no responsive documents from those sources and that there is a 0% related incidence of HVHF-related, adverse human-health impact(s) in that Region, including New York. (If data are ultimately received in response to our requests from the USA EPA Region 2, NY DOH, and/or NY DEC, an appropriate update to this publication will be released.)

The above-cited presence of methane in appreciable, if not remedial-level, amounts in 15% of 66 domestic water wells tested in southern New York, in the absence of current Marcellus, HVHF-type drilling and at least one mile from any existing well, suggests that background levels of methane in that geographic region are not associated with Marcellus, HVHF-type drilling. There may be parallel situations elsewhere in the Marcellus distribution, and those situations should be objectively addressed in terms of their ultimate source(s).

Resultant knowledge should be integrated into future regulations regarding HVHF-sourced drilling and production in the Marcellus.

Where there *is* current, Marcellus-related HVHF activity (Ohio, Pennsylvania, and West Virginia), and where roughly 11,560 total Marcellus wells have been drilled, there is a collective incidence of state-level environmental cases/claims of roughly 10% (969/10,000; see **Table 3**, above) in Pennsylvania (with a directly HVHF-related incidence of less than 2% [1.64-1.80%; see **Table 9**, above] there) and 0% in both Ohio and West Virginia (see **Table 9**, above), with a 0% incidence of documented, direct (adverse) human-health impact per the respective, two State Departments of Health (Pennsylvania has not responded, to date, to the author's inquiries) and two State Departments of Environmental Protection/Conservation (Pennsylvania, again, has not responded, to date, to the author's inquiries.) Despite the lack of direct response from the PA DEP to inquiries posed by this author, reliable secondary sources proved rewarding in terms of detailed, related content.

Regions 3 and 5 of the US EPA have each apparently experienced a 0% incidence of adverse human-health impact from the Marcellus HVHF activities, as reported by them as "no data responsive."

In summary, formal inquiries directed at nine (9) states' respective Departments of Health and Departments of Environmental Protection/Conservation, as well as at the US EPA covering the jurisdictions of interest, showed that there is no Marcellus/HVHF activity in five (5) of those states. Over the course of the past eight (9) months, the New York Department of Health has not provided a final, written response to similar inquiries. US EPA Region 2 would not release related information without cost-prohibitive fees, but there has been no HVHF-type Marcellus activity in either New York or New Jersey, which fall within that EPA jurisdiction. In the three (3) states where there has been extensive Marcellus drilling using HVHF, the respective State Departments of Health (excepting Pennsylvania, which did not respond to the inquiry) and Departments of Environmental Protection/Conservation have, to date, apparently experienced a 0% incidence of HVHF- related, adverse human-health impact, as have US EPA Regions 3 and 5. (The numerous, detailed, cited secondary sources of related information in Pennsylvania support this observation; see **Table 2**, above.) While there have been numerous cases of adverse environmental impact, apparently mainly from surface activities, in Pennsylvania – the relative hotbed of Marcellus HVHF-type drilling – as a result of Marcellus-related and/or "oil and gas operations," there has been an apparent 0% incidence of associated documented, adverse human-health impact.

According to these figures and to information made available to date by the cited agencies and reliable secondary sources cited above, the current incidence of acute, adverse human-health impact from the hydrofracturing process itself appears to be 0%.

These data, in turn, suggest that, to date, actual HVHF within the active Marcellus zones, according to the agencies and sources cited, has been, in very simple terms, 100% safe with respect to acute, adverse, human-health impacts. Where responses to related inquiries from this author remain unanswered, both secondary sources and extensive Internet searches have failed to reveal officially documented cases, or trends, of adverse human-health incidents related to HVHF in the Marcellus.

To put the latter into more scientific terms with regard to the statistics of rare events, it would be appropriate to invoke the mathematics of the Poisson distribution.[31,32] This methodology concludes that, when 0 (zero) "hits" in N attempts are experienced, this outcome is expected, to 95% or more probability, if the true mean is 3/N or less. For the wells in Pennsylvania, for example, the estimated true (population) mean is, to at least 95% probability, ≤ (less than or equal to) 3 adverse human-health incidents per 10,000 wells. In Pennsylvania, *none* of the above-cited incidents rose to the level of federally or state-documented adverse human-health outcomes, per primary and secondary sources on the matter, suggesting that the number of such outcomes, to date, is less than the mathematically/statistically predicted total.

Incidents of surface spills and surface phenomena directly or indirectly related to HVHF and/or HVHFF are common, as cited above and as might be expected where high volumes of concentrated chemicals and much traffic are found; however, such is not unique to hydrofracturing. A recent spill of a concentrated, potentially toxic chemical into the Elk River in Charleston, West Virginia, exemplifies that claim.[33]

These findings parallel the professional experiences of the author, who, for over five years, has practiced Occupational and Environmental Medicine in a relatively large healthcare center, in an extremely active area of Marcellus HVHF-type drilling and production, and who has professionally seen (aside from hundreds of work-related injuries and trauma to Marcellus workers themselves) *zero* confirmable cases of HVHF-related adverse human-health impact in local residents/citizens.

As additional regulatory controls of HVHF-related drilling for natural gas in the Marcellus are proposed and/or promulgated, these facts must be considered.

This is not to say that the cited lack of current, documented evidence of adverse human-health effects from HVHF in the Marcellus should be construed as a definitive, broad-based denial of such. The author, in fact, encourages ongoing and comprehensive searches for, and documentation of, HVHF-related, adverse human-health effects on residents in close, intermediate, and relatively distant proximities to HVHF activities/areas in the Marcellus. Such medical surveillance is a common-sense matter and would provide the "video" perspective in this matter, which may prove superior to the present "snapshot" one. A related database should be established, tracking all such claims and findings, by type, location, geology, and definitive etiology (where possible). Standardized, published, traceable standards and methodologies for such causality/causation determinations should be established and consistently applied if any sense is to be made, or conclusions drawn, from such a database. ■

6 REFERENCES

[1] www.geology.com/articles/Marcellus-shale.shtml

[2] www.bizjournals.com/denver/blog/earth_to_power/2013/04/us-natural-gas-reserves-hit- record.html?page=all

[3] www.eia.gov/tools/faqs/faq.cfm?id=52&t=8

[4] www.eia.gov/dnav/ng/ hist/n9140us2a.htm

[5] www.marcellusdrilling.com/2010/06/list-of-78-chemicals-used-in-hydraulic-fracturing-fluid-in-pennsylvania/

[6] www.anadarko.com/SiteCollectionDocuments/PDF/AXPC%20Real%20Facts%20about%20FracStim(2)012710.pdf

[7] www.endocrinedisruption.com/files/Pennsylvaniasummary4-20-09Final.pdf

[8] www.cdc.gov/healthywater / drinking/private/wells/

[9] www.eia.gov/pub/oil_gas/ petrosystem/us_table.html

[10] www.eia.gov/dnav/pet/hist/LeafHandler.ashx?n=PET&s=E_ERTW0_XWC0_NUS_C&f=A

[11] www.dec.ny.gov/ energy/46288.html#horizontal

[12] dl.dropboxusercontent.com/u /48182083/ScrantonTT_GW_RTK.csv

[13] Private communication, 8/12/13

[14] www2.epa.gov/aboutepa/our-mission-and-what-we-do

[15] www.serv.pa.gov/clientdocs/SERVPA_ops_manual_12_1_11.pdf

[16] www.depweb.state.pa.us/portal/server.pt/community/about_dep/13464

[17] www.odh.ohio.gov/rules/ombudsman/regulatoryombudsman.aspx

[18] www.epa.state.oh.us/about.aspx

[19] www.wvdhhr.org/bcf/aso/documents/ASOCodeofConduct.pdf

[20]www.wv.gov/pages/agency.aspx?newid=192

[21] www.dec.ny.gov/energy/46288.html#horizontal

[22] Private communication, New York Department of Environmental Conservation, 1/21/14

[23] www.stargazette.com/article/20090925/NEWS01/90925025/1117/news/Pa.+orders+a+halt+to+Cabot+gas+drilling+in+Dimock

[24] www.scientificamerican.com/article.cfm?id=natural-gas-make-water-burn

[25] www.forbes.com/sites/jonentine/2013/08/22/will-washington-politics-kill-the-us-energy-revival-and-shale-gas-revolution/2/

[26] Miller TL; Wolin MJ, de Macario EC, Macario AJ (1982). "Isolation of Methanobrevibacter smithii from human feces," Appl. Environ. Microbiol. 1982, 43(1):227–32. PMC 241804. PMID 6798932

[27] www.engineeringtoolbox.com/explosive-concentration-limits-d_423.html

[28] www.stargazette.com/viewart/20131217/NEWS01/312170062/Study-finds-methane-15-percent-upstate-N-Y-wells

[29] www.marcellus.psu.edu/index.php

[30] sites.nicholas.duke.edu/avnervengosh/duke-study-on-shale-gas-and-fracking/

[31] mathworld.wolfram.com/PoissonDistribution.html

[32] ncalculators.com/math-worksheets/poisson-distribution-example.htm

[33] www.cnn.com/2014/01/13/us/west-virigina-chemical-contamination/

BOARD OF TRUSTEES

FOUNDERS CIRCLE

ACSH STAFF

BOARD OF SCIENTIFIC AND POLICY ADVISORS

BOARD OF SCIENTIFIC AND POLICY ADVISORS (CONTINUED)

BOARD OF SCIENTIFIC AND POLICY ADVISORS (CONTINUED)

BOARD OF SCIENTIFIC AND POLICY ADVISORS (CONTINUED)

The opinions expressed in ACSH publications do not necessarily represent the views of all members of the ACSH Board of Trustees, Founders Circle and Board of Scientific and Policy Advisors, who all serve without compensation.

Made in United States
Cleveland, OH
05 November 2026

47649189R00026